OUT OF → THIS WORLD

VIRGINIA LOH-HAGAN

NOISES IN SPACE

45TH PARALLEL PRESS

Published in the United States of America
by Cherry Lake Publishing
Ann Arbor, Michigan
www.cherrylakepublishing.com

Reading Adviser: Marla Conn, MS, Ed.,
Literacy specialist, Read-Ability, Inc.
Book Designer: Jessica Rogner

Photo Credits: © Color4260/Shutterstock, cover, 1; © fizyk5/Shutterstock, 5; © Sergey Nivens/Shutterstock, 7; © NASA/NASA ID: iss060e015993, 8; © ALMA (ESO/NAOJ/NRAO), Kitayama et al., NASA/ESA Hubble Space Telescope, 11; © Alenavlad/Shutterstock, 12; © NASA/JPL, 15; © NASA/NASA ID: PIA10224, 16; © NASA Goddard/NASA ID: GSFC_20171208_Archive_e001322, 19; © NASA/NASA ID: KSC-99pc0136, 21; © NASA/NASA ID: 0300607, 23; © NASA Goddard/NASA ID: GSFC_20171208_Archive_e000078,25; Uploaded by: Saber1983/WikiMedia, 27; Grote Reber/WikiMedia/Public Domain, 29
Graphic Element Credits: © Trigubova Irina/Shutterstock

45th Parallel Press is an imprint of
Cherry Lake Publishing Group

Library of Congress Cataloging-in-Publication Data
Names: Loh-Hagan, Virginia, author. | Loh-Hagan, Virginia. Out of this world.
Title: Noises in space / by Virginia Loh-Hagan.
Description: Ann Arbor, Michigan : Cherry Lake Publishing, 2020 | Series: Out of this world | Includes bibliographical references and index.
Identifiers: LCCN 2020006917 (print) | LCCN 2020006918 (ebook) | ISBN 9781534169241 (hardcover) | ISBN 9781534170926 (paperback) | ISBN 9781534172760 (pdf) | ISBN 9781534174603 (ebook)
Subjects: LCSH: Cosmic noise--Juvenile literature. | Radio astronomy–Juvenile literature. | Sound–Juvenile literature.
Classification: LCC QC809.C6 L65 2020 (print) | LCC QC809.C6 (ebook) | DDC 551.5/276--dc23
LC record available at https://lccn.loc.gov/2020006917
LC ebook record available at https://lccn.loc.gov/2020006918

Printed in the United States of America | Corporate Graphics

TABLE OF CONTENTS

WHAT IS THE UNIVERSE?

The universe is huge. It's everything that exists. This includes planets, stars, and outer space. It includes living things on Earth.

The universe contains billions of galaxies. Galaxies are huge space collections. Galaxies are made up of billions of stars, gas, and dust. Galaxies include **solar** systems. Solar means sun. Earth is in the Milky Way galaxy. Galaxies spin in space. They spin very fast. There's a lot of space between stars and galaxies. This space is filled with dust, light, heat, and rays.

Before the birth of the universe, there was no time, space, or matter. Anything that takes up space is matter. Matter can exist in different states. The common states include solid, liquid, and gas. This is why things like air and smoke are considered matter. But the heat and light from a fire aren't matter. These don't take up space.

The universe hasn't always been the same size. It also hasn't always existed. Some scientists believe it began with a "big bang."

Over the years, the Milky Way has merged with other galaxies.

This is a **theory**. Theory means an idea. This theory explains how the universe was born. First, the universe was a super tiny blob, smaller than a pinhead! Then, that super tiny blob exploded. This happened 13.8 billion years ago. Next, energy spread out. Energy is made from matter. For example, the flames in a fire are matter. They take up space. But the heat you feel and the light you see from the flames are energy. Last, stars and planets formed. This all happened in less than a second.

Scientists think the universe is still expanding. Expanding means growing or spreading out. Scientists also think this expanding process is speeding up.

CHAPTER → ONE

IS THERE SOUND IN SPACE?

Sound travels in waves. It's different from light or heat waves. It makes particles **vibrate**. Particles are the smallest bits of matter. To vibrate is to move back and forth.

When something moves, it vibrates. It pushes air particles nearby. The particles push and pull each other. This motion is a sound wave. It has to travel through matter. Matter includes air, water, and solids. Matter acts as the **medium** of sound waves. Medium means a travel source.

We hear sound because sound waves vibrate particles in the air. These vibrations reach our ears. This is how sound is heard. Sound is vibrations in air. This is what happens on Earth.

The brain interprets vibrations as sounds.

But space is different. It's huge. Space objects are far from each other. There are millions of miles of empty space between objects. There's no medium for sound to travel through. There's no air, water, or solids. There are no particles to vibrate. So, many people think there's no sound in space.

Sounds in space have a hard time traveling to our ears. Most sounds die away. There's just too much space. Human ears aren't sensitive enough.

Scientists found a way to hear sounds in space. Radio waves can travel in space. Many space objects emit radio waves. Scientists use **satellites**. Satellites are machines in space. They have sensors. They "hear" radio waves. They pick up signals. Scientists change signals into sounds.

Astronauts can talk to each other in space. There's plenty of air in the spacecraft. Or they use their radios.

AMONG THE STARS: WOMEN IN SCIENCE

Dr. Wanda Diaz-Merced is an astronomer. Astronomy is the study of space. Diaz-Merced was born in Puerto Rico. She was born with diabetic retinopathy. She slowly lost her eyesight. She was completely blind by age 29. She came to the United States to study. She didn't let anything stop her from learning astronomy. She worked at NASA Goddard Space Flight Center. NASA stands for National Aeronautics and Space Administration. Aeronautics deals with aircraft. She did this in 2005. She found ways to turn large data sets into sounds. She did this to find patterns. She used information from telescopes. Diaz-Merced focused on the sounds. She has a TED Talk. Her TED Talk is called "How a blind astronomer found a way to hear the stars." She said, "I keep vigilant for the unexpected. Sensitivity to events increases when you use sound. In noisy data, those in our studies identify more peaks, dips, and pulses than just with sight." Pulses are beats or small bursts.

CHAPTER → TWO

WHAT DID THE "BIG BANG" SOUND LIKE?

After the big bang, the universe was still growing. For the first 700,000 years, matter was still **dense**. Dense means crowded and packed in. Sound waves could still travel through it. The universe cooled. Sound waves stretched. This made sounds sound deeper or lower. It continued for hundreds of years. The bang wasn't a loud bang. It was a long hum. The universe got bigger and bigger. Then the sound died away.

Dr. John G. Cramer is a scientist. He studies **physics**. Physics is the science of matter and energy. Cramer works at the University of Washington. He re-created the sound of the big bang. He used data from 2 satellites.

Cramer fed the data into a computer program. The computer changed the data into a sound. But the sound was really low. It was too low for humans to hear. Cramer had to boost the sound.

The big bang is known as the first sound in history.

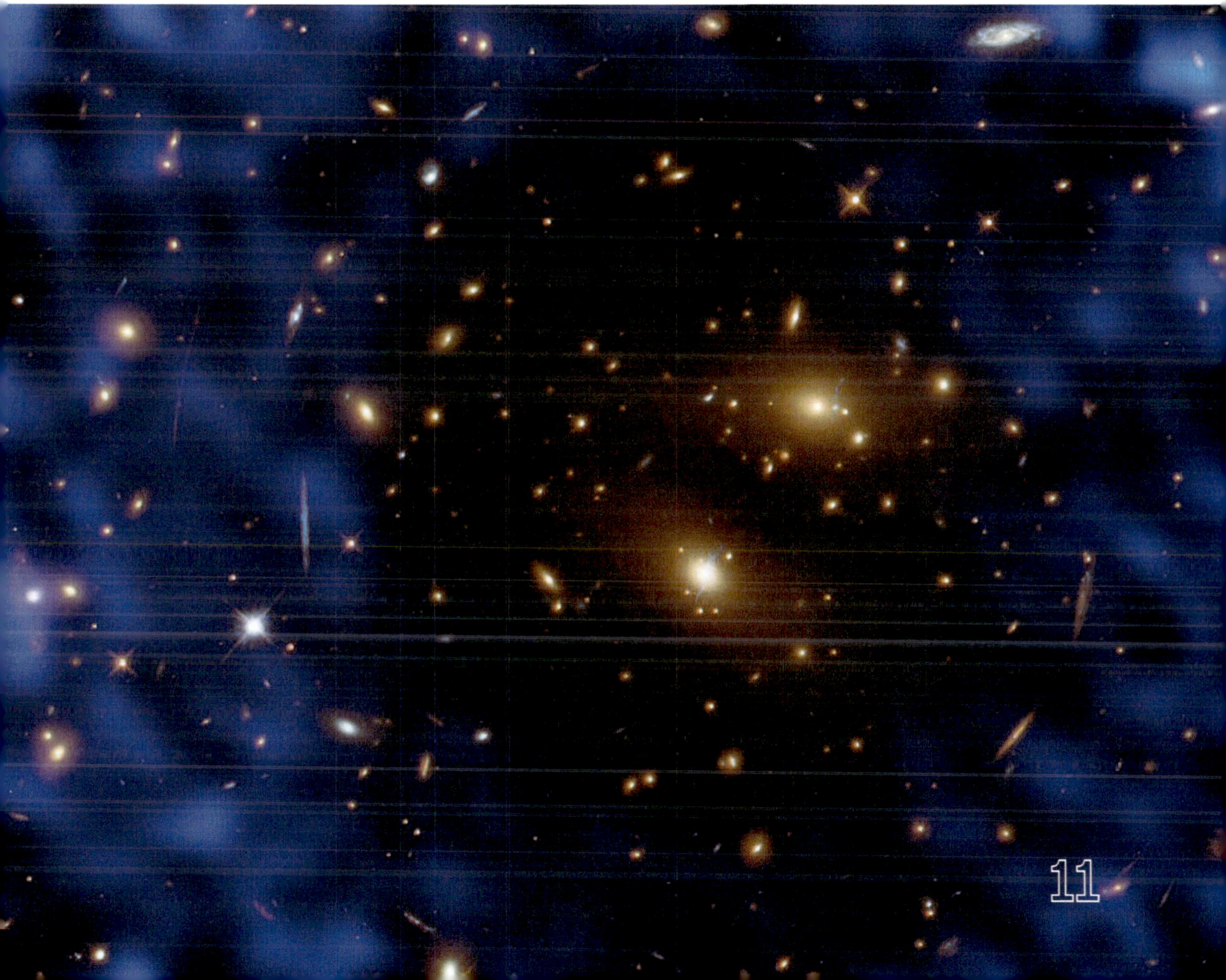

There are different descriptions of the sound. Cramer described it as a "large jet plane." He said the jet would be 100 feet (30 meters) off the ground. It would be flying over a house. He said the jet would be flying in the middle of the night.

The sound has also been described as a **robotic** humming. Robotic means machine-like. Some say it sounds like a "dying video game character." Others say it sounds like an "old computer powering down."

The big bang sound is like a bass instrument.

DOWN-TO-EARTH EXPERIMENT

Did you know you can "see" sound? Try out this experiment. Watch sound move.

Think like a scientist!

Materials:

- Plastic wrap
- Large glass bowl
- Tablespoon of uncooked rice
- Metal pan
- Metal spoon

Instructions:

1. Tear off a piece of plastic wrap. Cover the top of the bowl. Make sure the plastic wrap is tight.
2. Pour one spoonful of rice on top of the plastic wrap. Position the rice in the middle. Make sure the rice stays on top.
3. Hold a metal pan near the bowl.
4. Bang on the metal pan with a metal spoon. Watch what happens to the rice. Sound waves will reach the plastic wrap. The plastic wrap will vibrate. This vibration will cause the rice to move. It'll jump around. This is how you can "see" sound.

Sound waves move in all directions. This is why we can hear the banging sound.

This is why the banging sound moved the rice.

CHAPTER → THREE

WHAT ARE SOUNDS FROM JUPITER?

Jupiter is the fifth planet from the Sun. It's the largest planet in the Milky Way. It has 79 moons. It's noisy. Jupiter has interactions with its moons. It has many hot gases. All this causes intense radio storms. Jupiter emits strong radio waves. Its waves can be detected on Earth.

Juno is a spacecraft. Its goal is to study Jupiter. It has special tools. It has a sensor called Waves. Waves can find radio and **plasma** waves. Plasma is a very hot gas. Juno recorded waves. It recorded changes in the particles and fields.

Jupiter's magnetic fields are the strongest in our solar system.

It went to Jupiter in 2016. It crossed into Jupiter's **magnetic fields**. Magnetic fields are areas with moving electric charges. They're like belts around a planet. They protect planets from the Sun and from solar winds.

Juno moved from solar winds into magnetic fields. Solar winds blow past planets. They travel at 1 million miles (1.6 million kilometers) an hour. They hit up against Jupiter's magnetic fields. They slow down. They heat up. This causes **turbulence**. Turbulence is violent movement. Juno recorded this turbulence. It did this when it got into Jupiter's orbit. The sound was a **sonic boom**. Sonic booms are loud sounds. They sound like thunder.

Juno's sonic boom recording lasted 2 hours.

IT'S (ALMOST) ROCKET SCIENCE

NASA has a special space telescope. It's called Transiting Exoplanet Survey Satellite (TESS). TESS was launched on April 18, 2018. Its job is to search for the brightest stars near Earth. It has telescopes. It has detectors. It has special cameras. These cameras can see 85 percent of the sky. TESS can find sound waves in up to 1 million red giants. Red giants are big stars. They represent what our Sun would look like in 5 billion years. TESS finds these stars. Knowing more about stars means we can learn more about the planets that orbit them. TESS is called a "planet hunter." It finds exoplanets. Exoplanets are planets beyond our solar system. TESS is powerful. It's very sensitive. It picks up star movements. As of December 2019, it has found over 1,400 exoplanets. TESS orbits the Earth twice. It does this during the time the moon orbits once. Data is sent to Earth every 2 weeks.

CHAPTER → FOUR

WHAT HAPPENS WHEN A SPACECRAFT AND COMET MEET?

Comets are space balls with tails. They're made of ice and dust. Tempel 1 is a comet. It orbits the Sun every 5.5 years. It moves between Mars and Jupiter.

Stardust was a spacecraft. It was racing past Tempel 1. It was traveling fast. It did this in 2011. It flew through waves of Tempel 1's **debris**. Debris means scattered leftover pieces. Stardust got hit by comet dust particles. It got hit by small space rocks.

Icy dust balls that leave a comet keep breaking up in space.

Stardust had a Dust Flux Monitor. This is a special tool. It measures sound waves. It measures electrical pulses from dust impacts.

Stardust recorded its encounter with Tempel 1. The recording is 11 minutes long. It recorded over 5,000 impacts. The impacts sounded like bullets. They came in bursts. They came in clumps. Then, they crumbled.

Dr. Don Brownlee is a scientist. He works at the University of Washington. He's an expert on comets. He led the Stardust mission. He said, "The data indicate Stardust went through something similar to a B-17 bomber flying through **flak** in World War II." B-17 bombers are fighter planes. Flak occurs when guns from the ground fire at aircraft.

Stardust had a shield. The shield protected the spacecraft.

TETHER
YOUR
TOOLS

CHAPTER → FIVE

WHAT DO BLACK HOLES SOUND LIKE?

Black holes are dense areas in space. They make deep gravity sinks. Gravity is a force that attracts things to fall toward Earth's center. Black holes force things to fall into them. They pull in dust and gases from nearby stars. They pull so much that nothing escapes. Light can't even escape.

In 2003, scientists found sound waves from a **supermassive** black hole. Supermassive means really big. These are the largest black holes. They're equal to millions of suns. They were formed when star clusters collapsed together. Every large galaxy has a supermassive black hole at its center.

Black holes can't be seen.

These scientists used the Chandra X-ray Observatory. This is a special telescope. Telescopes are tools. They're used to seeing things that are far away. Chandra is in space. The scientists found the supermassive black hole in the Perseus cluster of galaxies. This cluster is about 250 million **light-years** from Earth. A light-year is the distance that light travels in a year. It's filled with gas. It has ripples in the gas. These ripples are proof of sound waves.

Scientists found that this supermassive black hole hums. The sound is the deepest note in the universe. It sounds like the B-flat note on a piano.

Scientists refer to this black hole as the "singing black hole."

CHAPTER → SIX

WHO FIRST STUDIED RADIO ASTRONOMY?

Karl Jansky lived from 1905 to 1950. He was born in Oklahoma. He studied physics. He went to the University of Wisconsin. He worked at Bell Telephone Labs.

Jansky built special tools. He built an **antenna**. An antenna is a machine. It sends and receives radio waves. Jansky's antenna turned in all directions. It was called "Jansky's merry-go-round."

He was looking for radio **static**. Static is a hissing noise. It's caused by electrical disturbances. It's called "white noise." Jansky made a surprise discovery. He heard radio waves coming from the Milky Way's center. He was the first to hear sounds

Jansky was a pioneer and inventor. He was curious about space. He wanted to learn more.

from space. He found this in 1931. He wanted to study this more. He became known as the Father of **Radio Astronomy**. Radio astronomy is a space science. It's the study of radio waves from space.

He studied static. He sorted static into 3 types. First, there was static caused by local thunderstorms. Second, there was static caused by distant thunderstorms. Third, there was mysterious static. Jansky studied the mysterious static. He figured out it came from the Milky Way.

Grote Reber lived from 1911 to 2002. He was a radio engineer. He built the world's first radio telescope. He did this in his backyard. He confirmed Jansky's findings.

Reber built the radio telescope in 1937.

GLOSSARY

ANTENNA (an-TEN-uh) machine that sends and receives radio waves

DEBRIS (duh-BREE) scattered leftover pieces

DENSE (DENS) crowded and packed in

FLAK (FLAK) when guns from the ground fire at aircraft

LIGHT-YEARS (LITE-yeerz) the distance that light travels in 1 year

MAGNETIC FIELDS (mag-NET-ik FEELDZ) areas around planets that have moving electric charges

MEDIUM (MEE-dee-uhm) travel source

PHYSICS (FIZ-iks) science of matter and energy

PLASMA (PLAZ-muh) very hot gas

RADIO ASTRONOMY (RAY-dee-oh uh-STRAH-nuh-mee) space science that studies radio waves from space

ROBOTIC (roh-BAH-tik) machine-like

SATELLITES (SAT-uh-lites) spacecraft probes that orbit in space

SOLAR (SOH-lur) relating to the sun

SONIC BOOM (SAH-nik BOOM) loud sound caused by aircraft traveling faster than the speed of sound

STATIC (STAT-ik) hiss caused by electrical disturbances

SUPERMASSIVE (soo-pur-MAS-iv) really big

THEORY (THEER-ee) an idea meant to explain something

TURBULENCE (TUR-byuh-luhns) violent movement in air or water

VIBRATE (VYE-brate) to move back and forth

FAR-OUT FACTS

- The Golden Record is a mixtape for space. It's 2 engraved golden records. They carry a message from Earth to any space aliens. They're in space inside the Voyager spaceships. The recording was made in 1977. It was made by scientists, artists, and engineers. It has greetings in different languages. It has messages from leaders. It has pictures. It has music. It has sounds from Earth.
- Chris Hadfield is a Canadian astronaut. He was commander of the International Space Station (ISS). He lived in the ISS for over a year. The ISS is in space. Hadfield recorded sounds from the ISS. He recorded the sounds of everyday life in space. One of his sounds was the toilet flushing. The flushing sounded like a turbo-charged vacuum cleaner.
- NASA collected sounds from space. They put together a playlist of sounds. They used satellites. They did this for Halloween. They collected "spooky sounds."

LEARN MORE

Space: *A Visual Encyclopedia.* London; New York, NY: DK Publishing, 2010.

Tyson, Neil deGrasse. *StarTalk: Everything You Ever Need to Know About Space Travel, Sci-Fi, the Human Race, the Universe, and Beyond.* Washington, DC: National Geographic, 2017.

Tyson, Neil deGrasse, with Gregory Mone. *Astrophysics for Young People in a Hurry.* New York, NY: Norton Young Readers, 2019.

INDEX

ABOUT THE AUTHOR

Dr. Virginia Loh-Hagan is an author, university professor, and former classroom teacher. She would love to play piano in space! She lives in San Diego, California, with her very tall husband and very naughty dogs. To learn more about her, visit www.virginialoh.com.